A CAD primer

By

Dr GOURI GOUTAM BORTHAKUR

MSc, PhD(Physics), Assistant Professor

Department of Physics, JIST, Cheniamguri, Sotai, Jorhat, Assam

A CAD Primer

Introduction

AutoCAD is a commercial software application from Autodesk Inc, USA. It is intended for 2D and 3D computer aided design (CAD) and drafting and is available since 1982 as a desktop application. Since 2010 onwards the software suite is made available as a mobile web- and cloud-based app, marketed as AutoCAD 360.The precursor of AutoCAD was derived from a program called Interact CAD, which begun in 1977 and released in 1979;also referred to in early Autodesk documents as MicroCAD. This software was written prior to Autodesk's (then Marinchip Software Partners) formation by Autodesk cofounder Mike Riddle.

The first version by Autodesk was demonstrated at the 1982 Comdex and was released in December of the same year.As of now, the 2016 release marked the 30th major release of AutoCAD for Windows. The 2014 release marked the fourth consecutive year of AutoCAD for Mac. Autodesk, AutoCAD, Civil 3D, DWF, DWG, and Top base are registered trademarks or trademarks of Autodesk, Inc., in the USA and/or other countries.

AutoCAD provides civil engineers, designers, surveyors, and drafters with a comprehensive, one stop Solution for the design, drafting, and management of a wide range of engineering projects. Using an industry-proven, dynamic engineering model, AutoCAD links design and production drafting, greatly reducing the time it takes to implement design changes and evaluate multiple scenarios. A change made in one place instantly updates an entire project, helping you complete projects faster, smarter, and more accurately. All team members work from the same consistent, up-to-date model, so they stay synchronized throughout all project phases.

Following are the top ten benefits of using autoCAD:

1. Provides a cost effective robust way to reduce errors with the dynamic engineering model.

2. Reduce purchase, deployment, and support costs with one complete solution.

3. Increase value to client by delivering more design alternatives in less time.

4. Take full advantage of existing AutoCAD skills to get up to speed quickly.

5. Create production sheets faster.

6. Be sure that production drafting is always in sync with your design.

7. Complete projects faster and reduce the chance of coordination errors using the
Civil 3D project environment.

8. Exploit data compatibility.

9. Build a foundation for your custom solution.

10. Clearly communicate design intent and complete final proposals with realistic 3D
rendering.

Applications of AutoCAD

1. AS AN ARCHITECTURAL PLANNING TOOL

A CAD Primer

AutoCAD provides its users with an intuitive user-interface that comes with built-in design lay-outs. These lay-outs include numerous templates that were specifically designed for architectural planning and building construction. So with an adequate knowledge of AutoCAD, anyone can take on projects that consist of designing architectural plans for construction purposes or building structures to be replicated in real-time. **Newer versions of AutoCAD also** provide architects and builders with the analytical tools needed to analyze a building's components and troubleshoot the stress and load levels of every support structure of a virtually designed building. This means that with AutoCAD, one can create an architectural plan, **design a building and** carry out specific analysis to know the buildings capacity and strengths before replicating it on a physical site.

2. AS AN ENGINEERING DRAFTING TOOL

The drawing of engineering components, infrastructure designs and analyzing HVAC systems plays a major role in most engineering –Civil, Mechanical, Systems and Electrical engineering— fields. And to do this while minimizing human errors, the use of a computer aided design application is recommended. AutoCAD is one of the recommended design software applications because it provides professionals in these niches with unique drafting tools that can be used to bring their engineering ideas to life with the accuracy they require. So in this stead, AutoCAD serves as software for designing mechanical components, analyzing electrical and piping systems and solving design issues that may arise.

3. AS A GRAPHIC DESIGN TOOL

Although there are arguable more advanced graphic design tools in the computer aided design community, AutoCAD's innate features that enable its users plan out architectural spaces, map them out and take advantage of the available space makes it a formidable design tool that can be used simultaneously with 3D Max, Maya and other design/animation tools when the need arises. AutoCAD supports the use of **DWG and DXF files which** can be exported from its interface to those of other advanced CAD applications to aid animation projects. This means that as an animator or graphics designer, you can take advantage of AutoCAD to create building structures and architectural plans which can now be exported for further design additions on advanced modeling applications.

4. IN 3D PRINTING

To **create a 3D printed object, an individual must go through these three processes**; choose the object, get a virtual 3D representation of the chosen object, and then feed the 3D printer this prototype to carry out the 3D printing process. It is clear to see—from the 3D printing process—that a model design tool or software definitely has a part to play and this is where AutoCAD comes in. With AutoCAD, 3D printing enthusiasts can create bespoke 3D models on its workspace for use in the 3D printing process. AutoCAD also ensures that file compatibility is not an issue for you can design your models on its interface and export your designs in the preferred '.stl' format which most 3D printers and slicing software functions with.

5. IN THE FASHION INDUSTRY

It is important to understand that the design pattern of every diamond, shining stone or jewelry you have had the pleasure of either wearing or viewing did not come by chance but from careful design considerations and plans. AutoCAD is a design software that comes with required tools needed to draft and design virtually anything of your choice, and the design of certain fashion items is no exception. This CAD software and intuitive interface, un-complicates the complications that comes with designing intricate shapes consisting of octagons, tetrahedrons and many more shapes you or I may have no knowledge of.

6. AS AN INDUSTRIAL DESIGN TOOL

The goal of every manufacturing and industrial organization is to make enough money to cover the cost accrued in producing any product and AutoCAD helps reduce that cost in many ways. With the use of its **CAD interface, industrialists can design working prototypes of virtually any object as well as test its functionality** during the design process. AutoCAD provides the tools to both design the initial prototype as well as tweak its ergonomics before the need to sink money into the actually production comes up. Subsequently, designers can also use the virtual prototype for presentations when advertising or seeking for funds.

Free licencing :

AutoCAD is licensed, for free, to qualifying students and teachers, with a 18-month renewable license available. The student version of AutoCAD is functionally identical to the full commercial version, with one exception: DWG files created or edited by a student version have an internal bitflag set (the "educational flag"). When such a DWG file is printed by any version of AutoCAD (commercial or student) older than AutoCAD 2014 SP1, the output includes a plot stamp / banner on all four sides. Objects created in the Student Version cannot be used for commercial use. Student Version objects "infect" a commercial version DWG file if it is imported in older versions than AutoCAD 2015.

An Industry standard:

The native file format of AutoCAD is *.dwg*. This and, to a lesser extent, its interchange file format *DXF*, have become de facto, if proprietary, standards for CAD data interoperability, particularly for 2D drawing exchange.AutoCAD has included support for .dwf, a format developed and promoted by Autodesk, for publishing CAD data.

Advantages of Using AutoCAD

1. A mid range Software
2. User Friendly
3. Fulfills most of the professional needs of a mechanical engineer
4. Lots of Time saving (but not always)

A CAD Primer

5. Paperless Environment

6. Ease of Accuracy — e.g. joining of lines head to head is not possible manually with desired accuracy

7. Easy to edit

8. Ease of Repetition / Addition

9. Reproduction of Drawing is easy, rapid and reliable in terms of dimensional accuracy and precision.

10. 2D and 3D production of drawings is very easy.

11. Zoom Command Facility allows us to see the minute details on larger scale view.

12. A facility of Modeling and making Parts library is available.

13. Parent – Child relationship allows multiple editing in single command, means if one part is edited all other linked parts are auto edited

14. Data handling, data storage and data sorting become very easy in soft form as compared to the conventional Printing and filing.

15. Data transmission to distant and remote areas is very easy and economical via emailing and link sharing.

16. Layers option allows us to hide or show some specific details of a complex assembly drawing for clear understanding.

17. There is no limitation of a Drawing size. You can draw a 2 light years long line through proper scale stetting.

18. Auto scale option is available. Full scale drawings can be formed.

A CAD Primer

19. Geometrical relationship can make quick mathematical operations easy, e. g. you can make an inscribed circle using dimensions of the triangle or polygon etc.

20. BOM or BOQ can be produced with accuracy

21. Mass, Area, Volume, Center of Gravity can be calculated

22. Auto Dimensioning – easy, accurate and rapidly done within no time.

23. Dimension Standards are strictly followed

24. Bullion operation is possible.

25. Copy, scale, Move, Stretch, rotate, Pattern commands make the task simple and easy

26. Draw Guiding through Osnap, Ortho etc.

27. Direct and Quick Drawing of Symmetrical objects. e.g circle, ellipse, polygon rectangles and triangles etc.

28. Geometrical Relationships are possible. Like two parallel lines, Tangent lines, perpendicular lines etc.

29. Fixed line thickness throughout the drawing. —— *remember that, "by default, the line thickness of auto Cad is zero"*

30. Fill, Hatch, Section lines, Chamfer and fillet commands make the software a blessing as these operations are very difficult in manual drafting.

31. Adding Text – Lettering – is a big facility

32. Images can be imported and traced out for digitization of conventional diagrams and drawings.

Disadvantages of Using AutoCAD
1. Expensive equipment is required. like Computer, Plotter, big screen monitor etc.
2. Registered software is expensive and it requires a heavy re-occurring annual fee.
3. Equipment is fragile, can be damaged drastically.
4. Continuous Updating of the equipment and software is needed.
5. Electricity is mandatory for running the equipment which is a big issue for third world countries like Pakistan which is facing a severe power crises these days.
6. Data storage is also fragile — high probability of data corruption, data loss is a big threat for all soft form data banks. To overcome this potential disaster, experts suggest "Mirroring of Data" that is storage of data on multi hard drives on multiple locations which involves a big capital allocation. Some disaster data recovery systems can also be implied to recover the lost data.
7. Piracy and hacking threats are always there when you use internet for data transfer or even storage of data on a computer which is connected to internet.
8. Special computer Skills are required.
9. Proper Maintenance, supervision and administration is required for computer networking.

AutoCAD Tutorial 1 : Getting started

A beautiful way of learning AutoCAD is to use to the command line. Although, the GUI design keeps changing, the shortcuts, command line inputs remain quite independent of the version change or upgrade. So when you practice learning CAD and you become conversant with the AutoCAD Command line, you will never have to unlearn anything and it will add to your feel and intuition about the product.

Few important tips :

Tip 1:To toggle visibility of the command line use Ctrl+9

Tip2 :Before starting the tutorial switch your workspace(WS) to "Autocad Classic" . This can be done in the following way

WSCURRENT→Autocad classic→

Tip3 :A good way is to save your drawing all through:

WSAUTOSAVE→0(Zero)

AutoCAD Task 1 Create space for the drawing

      a. Limits→Specify lower left corner→ Specify upper right corner→Z→ A→

AutoCAD Task 2 Create a slanting line

      a. Ortho off

      b. L→Specify first point→Specify nxt point →

AutoCAD Task 3   Create a H or V line

      a. Ortho on

      b. L→Specify first point→Specify nxt point →

AutoCAD Task 4 Create a rectangle

      a. Ortho on

      b. L→Specify first point→Specify nxt point →

      c. Repeat step b 2 more times→

      d. C{to close the curve instead of drawing the 4th line}

AutoCAD Task 5 Undo the last step

      a. U→

AutoCAD Task 6 Erase something

      a. E→Select objects(by clicking on the objects) cursor will be seen with a small "pigbox"→

AutoCAD Task 7   Orange square at the end of a line etc "Object snap"

      a. Press shift and Rt clk→ Select object snap from DDmenu

AutoCAD Task 8   Create a circle

There are two methods of creating a circle

(a)By specifying radius or diameter

      a. C→Speccify centre point→Specify radius →

      b. C→Speccify centre point→D(To specify diameter)-→<Dia> →

(b) By Specifying points lying on the circumference of the circle

      a. C→3P(To chose the three points on the circumference of the circle)→Chose points by mose click→

      b. C→2P(To chose the two points on the circumference of the circle)→Chose points by mose click→

      c. C→T(To chose the points on other object, to which the circle should be a tangent)→Chose points by mose click→

AutoCAD Task 9 Draw a tangent to a circle from a point

      a. L→ Specify first point

      b. Shift+ RtClk→Object snap→ tangent

Autocad Tutorial 2: The basics of draw command

AutoCAD Task 1 Offset an object (create parallel lines or concentric circles)

   a. O→Specify offset distance→Select object to offset(by Mouse Clk)→Speccify point on side to offset(Right or left etc of the original object)→

AutoCAD Task 2 Trim(Break the unwanted part)

   a. tr→ first Select Cutting edge(by mouse clk), then the objects→ clk on the part of the object to trim→

AutoCAD Task 3 Zoom to a particular section of the drawing

   a. Z→ Specify lower left corner→ Specify upper right corner→

AutoCAD Task 4 Measurement

   a. Di→specify firt point→ Specify next point or →

AutoCAD Task 5 List properties of an object

   a. Li→ Select objects

AutoCAD Task 6 Find area and perimeter of a drawing with vertices

A CAD Primer

    a. Area→Specify first corner point(click on all corners)→

AutoCAD Task 7 Area of a circle or a curve

    a. Area→O→ Select the object by MouseCLk→

AutoCAD Task 8 Area between two countours

    a. Area→ A(to add the outer area)→ select vertices/Object→

    b. Area→ S(subtract the inner area)→select vertices/Object→

    c. F2(For viewing details in AutoCAD text window)→

Autocad Tutorial 3 : More of draw commands

AutoCAD Task 1 Draw a polygon

There are three ways of creating a polygon:

(a) The edge length method: Here user will have to chose the starting point and the end point of an edge. The input is E . Here is how

  a. POL→<no of sides>→E(To choose the edge length method)→Specify the first pint → Specify the second end point→

(b) The inscribed circle method : In this method, the user choses the centre of the poygon and the distance of the corner to the centre of the polygon. The keyword is I

  a. POL→<no of sides>→<centre>→I(To chose the inscribed circle method)→ specify radius→

(c)The Circumscribed method : Anoother way of creating a polygon is to specify the distance between the centre of the polygon and the midpoint of any one edge. The trigger is C.

A CAD Primer

POL→<no of sides>→<centre>→C(To chose the circumscribed circle method)→ specify radius→

AutoCAD Task 2 Create a rectangle

There are Two ways of creating a rectangle

(a) By Specifying the diagonally opposite corners

Rec→Specify first corner point→ Specify the other corner point→

(b) By specifying the area/dimension/rotation according to specifications

Rec→Specify first corner point→A(to specify area)→Specify other corer point→

Rec→Specify first corner point→D(to specify length and breadth)→<L→<B>→ Specify other corer point→

Rec→Specify first corner point→ R(To secify the inclination of the longest edge to the + xaxis)→<Angle>→Area/Dimension→ Specify other corer point→

AutoCAD Task 3 Mark something in your design for later revision (to create a revision cloud around those objects to be revised later)

a. Revcloud→

AutoCAD Task 4 Create an arc

a. A(to create an arc)→start point, centre point, End point

b. A→c(to specify the centre)→ start point, end point

AutoCAD Task 5 Create a spline

a. SPL(for spline)→specify points in zig-zag mode by mouse click→

b. Click on the spline→edit/stretch/refine/remove vertex(to adjust the shape of the spline)

AutoCAD Task 6 Create an ellipse

a. El→A(to crete an elliptical arc)→specify major axis diameter, minor axis radius, start point and end point of the erc→

      b. El→C(to specify the centre of the ellipse)→ specify semi-major axis and semi minor axis by mouse click→

AutoCAD Task 7 Insert points

      a. Po(to insert points)→Mouse clk at desired location→Alt+o→Point Styles→ Chose→

AutoCAD Task 8 Create a donut or a filled circle

      a. Do(For Donut)→Specify inside diameter→Specify outside diameter→ Mouse clk centre→

      b. For Filled circle, give inside diameter zero

A CAD Primer

Autocad Tutorial 4 : Modify Commands

AutoCAD Task 1 Copy an object

    a. Note :An important concept in autocad is the concept of base point. A base point is a reference point, with respect to which, the desired command or instruction will be executed in autocad.

    b. Method1 : CO(to copy something)→select objects by mouse clk→specify base point in the object→Specify second point(s)(the points where the copied object is to be placed)→

    c. Method2: CO(to copy something)→select objects by mouse clk→specify base point in the object→D(to displace the object by a certain distance)→

AutoCAD Task 2 To make a mirror image of an object

    a. MI(to enable the mirror)→select object→Specify first point of the mirror line→ Sepcify second point of the mirror line→Erase source object→N(if you don't want to earse the originall object)→

AutoCAD Task 3 Make an array

a. There are two ways to create an array, the rectangular array, and the polar array which can be created as below in the array creation default settings

b. For rectangular array:AR(to enable the array)→select the object to be arrayed→R(to create a rectangular array)→

c. To change the settings of the default rectangular array thus created,enter

    i. S(for changing the spacing)→(enter positive number if you want the array to be on the rght and above of the source object, and enter negative number if you want the array to be on the left and below the source object)

    ii. COL(for changing the number of columns)→

        iii. R(For changing the number of Rows)→

d. For polar array:Method 2 AR(to enable the array)→select the object to be arrayed→PO(to create a polar array)→ A(to create a polar array around an Axis)→Specify the endpoints of the line →

e. For polar array:Method 1 AR(to enable the array)→select the object to be arrayed→PO(to create a polar array)→ B(to create a polar array around the ccentre of a circular arc)→

f. To change the settings of the default polar array thus created,enter

        i. I(for number of Items)→

        ii. F(for changing Angular width of the region to be filled with the array)→

        iii. ROW(for changing Rows of the array)→ etc

AutoCAD Task 4 Move an object in the drawing

a. MOVE→ Select object→Specify base point(point in the object to be moved wrt which the move command will be executed)→specify second point(point to which the object would be moved)→

AutoCAD Task 5 Rotate an object

a. RO(to rotate)→Select objects→Specify base point(the point around which, the object will be rotated)→Specify rotation angle→

AutoCAD Task 6 Stretch an object or a part of it

a. In AutoCAD, there are two kinds of select window:

   i. the blue select(when you move the mouse from left to right) window, or the normal window;

   ii. and the green select(when you move the mouse from right to left) window or the crossing window

   iii. The objects for which the length is to be changed should be half selected by the crossing window, while the objects, which are to be moved by the stretch command should be completely selected by the crossing window.

b. STRETCH→Select object(use crossing window)→ Sepcify base point→ Specify end point/ Displacement

AutoCAD Task 7 Exteend and object or a part of it

a. EX(to extend)→ Select boundary edges(upto which the object selected is to be extended)→Select Objects→

AutoCAD Task 8 Break an Object between two points

a. BR(ie to break)→Select objects→F(to specify thee first break point)→ Specify second break point→

b. The first breakpoint, by default is the point on the object, where you clicked first to select the object.

AutoCAD Task 9 Insert the breakline symbol

    a. BREAKLINE→Specify first point, secondpoint and the midpoint of the breakline symbol in order→

AutoCAD Task 10   Join Multiple objects

    a. JOIN→Click on objects to join→

AutoCAD Task 11   Apply Chamfer at a corner

    a. CHAMFER→D(to specify chamfer distances)→ Specify first and second chamfer distance→ Select first and the second line →

AutoCAD Task 12   Apply a Fillet at a corner

    a. FILLET→ R(to specify the fillet radius)→ Select first and the second line →

Autocad Tutorial 5 : Layers and Hatches

AutoCAD Task 1 To   create layers and edit layer properties

    a. LA(to open the layer Properties manager)→Click anywhere on the dialoguebox→Alt+N(for new layer)→Give a meaningful name→edit line color, linetype, linewidth for the layer→Repeat the procedure for as many layers you need

AutoCAD Task 2 To switch between different layers

    a. There will be a "Layer properties manager(green contour)" and a layer control toolbar just above your workspace, where you will see the label of the layer you are currently working on.Click on the layer name and select the desired layer from the "Layer control(red contour)" dropdown menu.

AutoCAD Task 3 Make a layer invisible

a. Click the bulb icon near the layer control bar. If it turns yellow, the layer is now visible; if the bulb turns gray on click, the layer becomes invisible

AutoCAD Task 4 Make a layer unmodifyable

a. Click on the lock icon on layer control bar. If the lock gets locked, the corresponding layer os unmodifyable; On clicking again, the lock gets opened and one can then moify the contents of the layer.

AutoCAD Task 5 : Modify the line type scale (for example, increase the gap between dots in a dotted line, or say, change the gaps between the .- . lines)

a. Select the layer in which the object is located, the line type of which is to be modified.

b. LTS→ Enter scale factor→

AutoCAD Task 6 Copy the line type properties of any object(called the source object) to another object(called the destination object) drawn with a different line type

    a. MA(Match Properties)→ Select source object→ Select Destination object→

AutoCAD Task 7 Move the object in one layer to another layer

AutoCAD Task 8 Select object by mouse click→Select the destination layer from the layer control toolbar→

AutoCAD Task 9 Apply a hatch

    a. H(to apply hatch)→Hatch and gradient dialogue Box appears→Select swatch,angle, scale and gradient if any→

    b. If the object to which the hatch is to be applied does not have a closed boundary, then in the boundaries part of the Diag Box, select **Add:Select objects → Green select object→**

    c. If the object to which the hatch is to be applied have a closed boundary, then in the boundaries part of the Diag Box, select **Add:Pick points→ Click inside the object to be hatched→**

d. In hatching <u>multiple objects,</u> apply hatch simultaneously, and **click "create separate hatches"**

e. In Hatching <u>nested objects,</u> one needs to play with the **Islands section of the hatch and gradient dialogue box**(if islands section **is not visible, click on the > icon at the bottom right** of the dialogue)